Distinctions in Nature

# Vertebrates and Invertebrates Explained

Shirley Duke

Cavendish Square
New York

Published in 2017 by Cavendish Square Publishing, LLC
243 5th Avenue, Suite 136, New York, NY 10016



First Edition



Library of Congress Cataloging-in-Publication Data

Names: Duke, Shirley Smith, author.
Title: Vertebrates and invertebrates explained / Shirley Duke.
Description: New York : Cavendish Square Publishing, [2017] |
Series: Distinctions in nature | Includes bibliographical references and index.
Identifiers: LCCN 2016008999 (print) | LCCN 2016010773 (ebook) |
ISBN 9781502617798 (pbk.) | ISBN 9781502620125 (library bound) |
ISBN 9781502617613 (6 pack) | ISBN 9781502617538 (ebook)
Subjects: LCSH: Animals–Classification–Juvenile literature. |
Anatomy–Juvenile literature. | Vertebrates–Juvenile literature. |
Invertebrates–Juvenile literature.
Classification: LCC QL351 .D85 2017 (print) | LCC QL351 (ebook) |
DDC 590.1/2–dc23
LC record available at http://lccn.loc.gov/2016008999

Editorial Director: David McNamara
Editor: Kelly Spence
Copy Editor: Nathan Heidelberger
Art Director: Jeffrey Talbot
Designer: Stephanie Flecha
Production Assistant: Karol Szymczuk
Photo Research: J8 Media

The photographs in this book are used by permission and through the courtesy of: File:2010-kodiak-bear-1.jpg/Yathin S. Krishnappa/Wikimedia Commons, cover left; Fleetham Dave/Perspectives/Getty Images, cover right; dpa picture alliance archive/Alamy Stock Photo, 4; File:Vertebrates.png/Wikimedia Commons, 6; Rich Carey/Shutterstock.com, 7; David Hosking/Alamy Stock Photo, 8; George Grall/National Geographic/Getty Images, 10; Maxfromhell/Shutterstock.com, 11 top; musk/Alamy Stock Photo, 11 bottom left; Dorling Kindersley/Getty Images, 11 bottom right; Image Source/Getty Images, 12; Andrii Vodolazhskyi/Shutterstock.com, 14 top; BSIP/Universal Images Group/Getty Images, 14 bottom; Feargus Cooney/Lonely Planet Images/Getty Images, 15; U.S. National Oceanic and Atmospheric Administration, 16 top; scorpion56/iStockphoto, 16 bottom; Nancy Nehring/iStockphoto, 17; aquapix/Shutterstock.com, 18; Dennis W. Donahue/Shutterstock.com, 20 top; Rasmus Loeth Petersen/Alamy Stock Photo, 20 bottom; Jay Boucher/Shutterstock.com, 22 top; encikAn/Shutterstock.com, 22 bottom; Tom McHugh/Science Source/Getty Images, 24; AppStock/Shutterstock.com, 26; Dennis Sabo/Shutterstock.com, 27 left; Christian Darkin/Science Source, 27 right.

Printed in the United States of America

# Contents

A giraffe's long neck has the same number of small bones as a human neck. However, each bone is much longer.

# Introduction: All Animals

The animal kingdom is made up of all sorts of creatures. Some are tiny, like bacteria. Others are large, like an elephant. A **skeleton** is the structure that gives an animal its shape and allows it to move. Not all skeletons are made of bone, or found inside an animal.

## Classifying Them All

**Taxonomy** is the science of classifying things. Scientists put living things into groups that are alike. The two main animal groups are vertebrates and invertebrates. Each group is then divided into smaller groups. These groups are based on shared **traits**, or characteristics.

The pictures below show examples from the five main vertebrate groups. The fire salamander is an amphibian, while the saltwater crocodile is a reptile. The southern cassowary is a bird, the giant elephant shrew is a mammal, and the ocean sunfish is a fish.

Vertebrates, which are animals with backbones, live in every part of the world. Fish, **amphibians**, **reptiles**, birds, and **mammals** are all part of this group. As mammals, humans are vertebrates.

## Invertebrates

Invertebrates are animals without backbones. They are found in all habitats. They include a **diverse**, or wide, range of animals. There are over thirty groups of invertebrates. These include sponges, corals, worms, and spiders. Crabs, mussels, and octopi are also invertebrates. Insects make up the largest group. All

## Zoom In

Many more invertebrates live on Earth than vertebrates. From 1 to 5 percent of all animals on Earth are vertebrates. Invertebrates make up 95 to 99 percent of all animals.

An octopus makes its home in the cracks and openings of a coral reef.

invertebrates share some things in common. Each invertebrate group also has its own special traits.

A lamprey in an ancient vertebrate. This species of fish is believed to be at least three hundred million years old.

# 1 From the Beginning

Invertebrates are less developed than vertebrates. Vertebrates **evolved**, or changed over time, from invertebrates. As they developed, they learned how to swim, walk, slither, and fly.

## In the Water

The early vertebrates had no jaws. Some ancient vertebrates still exist today. Hagfish and lampreys do not have jaws. Instead, they suck or bite on food using their teeth. Over time, fish developed the first jaws. These animals live in water and move with

Frogs are amphibians. As tadpoles, over time they go through a change in form. Their gills shrink and they grow lungs. They also lose their long tails and grow legs.

fins. They are covered in scales. Fish breathe through **gills**, small openings that take in oxygen from water.

## On the Land

Animals continued to evolve. The next group to develop were the amphibians. Amphibians, like salamanders and frogs, live part of their lives in water and part on land. They have moist skin and lay their eggs in the water. Reptiles came next. Reptiles live on land and lay eggs there. They are covered in rough scales. Snakes, lizards, and turtles are reptiles.

## Zoom In

All invertebrates are ectothermic, or cold-blooded. This means they cannot create heat within their bodies. Some vertebrates, like fish, reptiles, and amphibians, are also cold-blooded. Only birds and mammals are warm-blooded.

Birds evolved from the dinosaurs, an early type of reptile. Mammals arrived next. They have fur and give birth to live young. The larger brains of mammals makes them the most advanced animals.

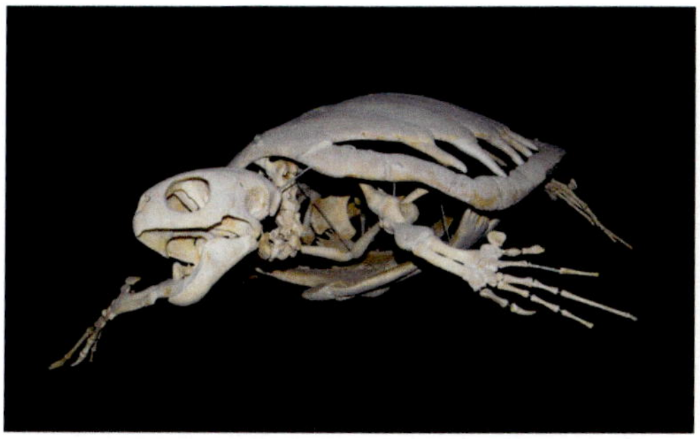

This turtle skeleton's backbone is hidden beneath its shell.

A chimp's brain is among the most developed of all vertebrates.

The bones in an elephant's skeleton are large and thick to support the animal's weight. The vertebrae are not flexible and have flattened discs to help support the head and trunk.

# 2 Comparing Vertebrates and Invertebrates

All vertebrates have a **spine**, or backbone. The spine is made up of several small bones that join together. Each tiny bone is called a **vertebra**. Each vertebra is stacked on another and helps protect the **spinal cord**. With a strong, flexible spine, vertebrates are able to grow to large sizes. The skeleton grows with the animal as it gets bigger. Most have two to four limbs, such as fins, legs, or wings.

## Zoom In

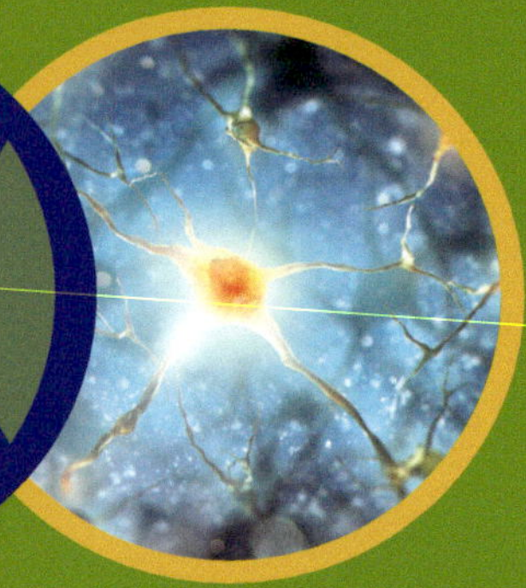

Vertebrates tend to be much smarter than invertebrates. Vertebrates have well-developed nervous systems. Their brains and spinal cords connect to nerves running throughout their bodies. Their backbones protect their spinal cords. A hard skull protects these animals' more developed brains.

## Strong Sacs and Body Armor

Invertebrates are generally smaller than vertebrates. They do not have any bones to support their bodies. Some, like jellyfish and worms, have sac-like skeletons filled with

Hens, like other vertebrates, have a backbone. The backbone protects the spinal cord running from the base of the brain down through the animal's body.

As a worm moves along, tiny bristles on the outside of its body help the animal to grip the ground as its muscles contract.

fluid. To move, their muscles **contract**, or shorten. This causes the animal's shape to change and allows for movement.

Other invertebrates have a hard outer shell made up of different jointed sections. This is called an **exoskeleton**.

## Zoom In

Arthropods that live in water are able to grow larger than those on land. This is because the water makes their exoskeleton feel lighter.

It protects the animal's soft body like a suit of armor. Insects, spiders, and some shellfish all have exoskeletons. Members of this group of invertebrates are called **arthropods**. As an arthropod grows, it sheds its exoskeleton. Then it grows a new one.

The wood louse, also known as a pillbug or roly-poly, is able to roll into a ball to protect itself because of its segmented body shape.

## Mirror Image

If you divide a vertebrate's body in half, each half will be identical. This is called **bilateral symmetry**. Some

The four-lined whiptail lizard displays bilateral symmetry.

This necklace sea star shows radial symmetry. It moves slowly along the ocean floor using tiny tubes on the underside of its arms.

invertebrates also have bilateral symmetry. Others have **radial symmetry**. This is an equal, rounded form growing out from a central point.

## Spiky Skin

Invertebrates like sea stars, sea urchins, and sea cucumbers are called **echinoderms**. Their skin is covered with sharp spikes. They are more closely related to vertebrates than any other invertebrates. As young, they have bilateral symmetry. After going through a change in form, they become radially symmetric.

1. A sungazer lizard basks in the sun to warm up.

2. The common octopus hunts at night. It uses its body coloring to blend in with rocks and coral.

# 3 Be a Body Detective

Use what you have learned to figure out which animals (*left*) are vertebrates and which are invertebrates.

1. This creature is ectothermic. It lies in the sun to warm up. As it grows, its bones grow, too. Is this a vertebrate or an invertebrate?

2. This animal has eight long legs. It is one of the most intelligent animals in this group. Do you think it is a vertebrate or an invertebrate?

3. Fast-moving water can carry sand dollars along with it, so they grow heavier skeletons. Young sand dollars may swallow sand to make themselves heavier and help hold them down.

4. Crickets eat fruit, nectar, insects, leaves, and dead animals.

3. This creature lives in the ocean. It burrows into the sand and is covered in tiny spikes. Is it a vertebrate or an invertebrate?

4. This animal's body is covered in a hard shell. When it grows too big for its shell, it will shed it. Then it will grow a new one. Is this animal an invertebrate or a vertebrate?

## Answer Key:

1. This sungazer lizard is a vertebrate. It has a backbone and belongs to the reptile group.
2. This octopus is an invertebrate. It does not have a backbone. It is one of the smartest members of the invertebrate family.
3. This sand dollar is an invertebrate and an echinoderm.
4. This cricket is an invertebrate. With its hard exoskeleton, it belongs to the arthropod family.

This ancient fish, a hagfish, has slimy, soft, loose skin and no jaw. The slime helps the hagfish get away from animals trying to eat it.

# 4 Rule Breakers

There are many unique animals on Earth. Some are very different than other members of their group.

## Not a Real Backbone

Hagfish are an ancient **species** of fish. While they have a skull, they do not have a true backbone. Instead, they have a skeletal structure made of **cartilage**. The skeletons of sharks and rays are also made of cartilage.

Special scales on a snake's belly line up with its ribs. The animal's backbone is flexible and its many vertebrae allow it to move in different directions by using its muscles to push against rocks, sticks, or the ground.

## Limbless Vertebrates

Not all vertebrates have limbs. Snakes are reptiles that have a skull and a backbone running through their bodies. Snakes have hundreds of vertebrae that twist and turn as they slither along the ground. Some amphibians, like certain species of salamander, are also limbless.

## Unique Sponges

Almost all invertebrates are symmetrical. However, sponges have no symmetry. These ancient animals mostly do not have **tissues** or **organs**, either. As

Sponges, like these finger sponges, fasten onto rocks with special body parts that look like roots.

The giant squid glides through the water by pumping water through its body. Extra fins on its head help the animal to steer.

they grow, sponges stay in one place. They are found in both fresh and salt water.

## One Big Invertebrate

Most invertebrates are small. Many of them can't support a large body. The giant squid is the largest invertebrate on the planet. The largest giant squid discovered by scientists measured 59 feet (18 meters) long. It weighed almost 1 ton (900 kilograms)!

## Glossary

**amphibians** Cold-blooded animals with smooth, moist skin that lay eggs in water and go through two stages of development, one in water and one on land.

**arthropods** The group of invertebrates that have jointed legs, segmented bodies, and exoskeletons.

**bilateral symmetry** When the two parts of an animal make a mirror image if divided down the center.

**cartilage** A hard, flexible material made up of tissue.

**contract** To shorten.

**diverse** Very different; having lots of variety.

**echinoderms** The invertebrate group that includes spiny creatures like sea urchins and sea stars.

**ectothermic** Describing an animal that depends on its surroundings to regulate its body temperature.

**evolved** Developed slowly as a species over time to a more complex form.

**exoskeleton** A hard, outer shell.

**gills** Structures that take oxygen from water, allowing animals to breathe underwater.

**mammals** Animals with fur that give birth to live young and feed their young milk from their bodies.

**nervous systems** Body systems made up of the brain, spinal cord, and the nerves they connect to.

**organs** Groups of like tissues working together on a job in the body of an animal.

**radial symmetry** An even, rounded form around a central core.

**reptiles** Cold-blooded animals that lay eggs and are covered with scales.

**skeleton** The part of the body that gives structure to an animal.

**species** Kinds of animals with similar traits.

**spinal cord** The large nerve running down an opening in the vertebrae to other nerves.

**spine** The backbone.

**taxonomy** A method of grouping or classification.

**tissues** A group of like cells working together on a certain job.

**traits** Characteristics of living things.

**vertebra** A single bone in the spinal column.

# Find Out More

## Books

Arnold, Caroline. *Living Fossils: Clues to the Past.* Watertown, MA: Charlesbridge Publishing, 2016.

Harris, Tim. *The Animal Kingdom*. Science Q & A. New York: Cavendish Square Publishing, 2015.

Kalman, Bobbie. *What Is a Vertebrate?*. Big Science Ideas. New York: Crabtree Publishing, 2007.

## Websites

DK Findout! What Is a Vertebrate?

www.dkfindout.com/us/animals-and-nature/animal-kingdom/what-is-vertebrate

Learn all about vertebrates and read about different animals in this group.

DK Findout! What Is an Invertebrate?

www.dkfindout.com/us/animals-and-nature/invertebrates

Explore the diverse world of invertebrates.

# Index

Page numbers in **boldface** are illustrations.

## About the Author

**Shirley Duke** has written more than fifty books about science and nature, but writing about life science is her specialty. She studied vertebrates and invertebrates and these animals are her favorite part of science. She sees many vertebrate animals near her home in the Jemez Mountains of New Mexico. She's observed elk, deer, cougars, bears, coyotes, badgers, raccoons, chipmunks, squirrels, ground squirrels, and birds. Most of these she hadn't seen until she moved to the mountains.